Acknowledgments

We would like to first thank our publisher for the opportunity to share the medium of soapstone carving in another great book and thank you to Hobby Lobby for supplying the materials and supplies. We also thank our artist, Dawn Hartwig, whose years of experience and talent show in even the most basic of projects and we are just thrilled to share with you her techniques and approach. Thank you, Dawn, for working with us. We would like to thank our dear friends and family who have encouraged us through this process. We hope that you will find as much inspiration in carving the inspirational sitting bear as we have. Happy Carving!

In loving memory of Phil Arnott

Contents

1

Carving a Sitting Bear in Soapstone

Tasha Unninayar and Lynn Bartlett
Artwork by Dawn Hartwig

Schiffer Publishing Ltd

4880 Lower Valley Road • Atglen, PA 19310

Introduction to Soapstone Sculpting, 978-0-7643-3781-9, $29.99
Carving a Coyote in Soapstone, 978-0-7643-4009-3, $16.99
Carving a Bear in Soapstone, 978-0-7643-4084-0, $16.99

Other Schiffer Books on Related Subjects:
Carving Bears and Bunnies, 0-8874-0267-4, $9.95
Bears to Carve with Dale Power, 0-8874-0-719-6, $12.95

Library of Congress Control Number: 2012944717

Designed by Mark David Bowyer
Type set in SquireD / Humanist521 BT

ISBN: 978-0-7643-4148-9
Printed in China

Published by Schiffer Publishing, Ltd.
4880 Lower Valley Road
Atglen, PA 19310
Phone: (610) 593-1777; Fax: (610) 593-2002
E-mail: Info@schifferbooks.com

For the largest selection of fine reference books on this and related
subjects, please visit our website at:
www.schifferbooks.com.
You may also write for a free catalog.

This book may be purchased from the publisher.
Please try your bookstore first.

We are always looking for people to write books on new and related
subjects. If you have an idea for a book, please contact us at:
proposals@schifferbooks.com

Schiffer Books are available at special discounts for bulk purchases for
sales promotions or premiums. Special editions, including personalized
covers, corporate imprints, and excerpts can be created in large quantities
for special needs. For more information contact the publisher.

In Europe, Schiffer books are distributed by:
Bushwood Books
6 Marksbury Ave.
Kew Gardens
Surrey TW9 4JF England
Phone: 44 (0) 20 8392 8585; Fax: 44 (0) 20 8392 9876
E-mail: info@bushwoodbooks.co.uk
Website: www.bushwoodbooks.co.uk

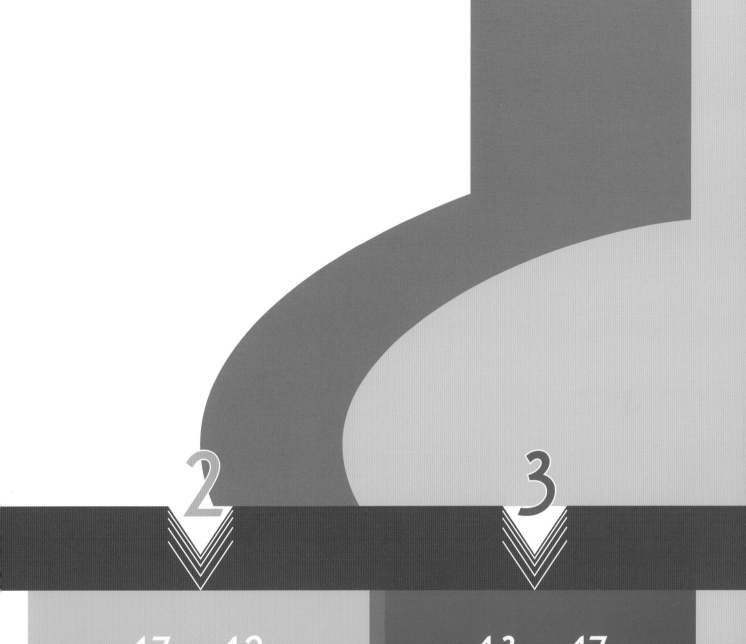

2

3

A Brief Introduction to Soapstone

There are many materials available for sculptors to create with such as wood, clay, marble, alabaster, and soapstone. Soapstone has always been a favorite because of its softness and ease to sculpt and manipulate. New sculptors of any age and skill level can easily accomplish fun projects with soapstone. This positive reinforcement encourages them to continue creating and learning this craft. The art of carving stone gives you a special feeling of creating something that will last forever and maybe even be a part of history. Your piece may even become a family heirloom. There are many techniques that you can learn that include the use of electric tools, found at your local hardware store, to ancient techniques that have been passed down for generations. The best part is that you don't need much to accomplish your carving. You can do it at home, in the garage, in a classroom, or any comfortable space of your choosing.

Raw soapstone.

White marble in raw form, wood block, golden soapstone in raw form, and clay formed into a seal.

How to Carve a Sitting Bear

History of the Sitting Bear

The bear has been the symbol of strength in Native American Indian cultures. The bear often referred to as the "great spirit" has been depicted in many forms, one of which is this rounded bear. The great bear spirit inspires strength, hard work, protection and love with its gentle form. This round bear is a simplified version of the traditional spirit bear.

Photo of a traditional Native American Spirit Bear.

Preparing Your Space

Hopefully you are lucky enough to be able to work out of doors on your projects leaving the debris easy to clean up. If this is the case find a table that you can work on (one that does not strain your back). Using lots of newspaper, cover your table in layers that you can remove as dust accumulates. If you put a wet towel under your soapstone project, you can trap the dust and discard it as you go by rinsing out the towel. You can create a "sandbag" out of an old pair of blue jeans. Cut a leg into a section and stitch one end. Fill the leg with silica sand (do not over stuff the leg) and stitch the other end — bingo you have a wonderful place to work your stone.

If you are forced to work indoors, you can use a kitty litter box, or any small box, and fill it halfway with silica sand (purchased at the hardware store — they refer to it as Play Sand). This box will now act as both a cushion for your piece to be worked on, as well as a collector for the dust.

Supplies You Will Need

For your bear, you will need a piece of soapstone; in this case, we used a precut stone of the bear design. (You can purchase the stone precut at most suppliers or request your supplier to cut it to size, and/or cut it yourself). You can also use a raw piece of stone or a block of stone and cut it down to the basic shape.

In this project, we used a rasp, a four-sided file, and a flat chisel. The carving could be accomplished with a mere rasp, but this process would take quite a bit more time and effort. You will also need a pencil, sandpaper (beginning with 80 grit, 120 grit and 220 grit), and some olive oil, or spray lacquer, for the finish. Optional supplies are a plastic mat to set the piece on, a plastic box, newspaper or old towel to catch the debris from the filing, and steel wool.

Finished Round Bear.

Raw soapstone, blocked stone, and a precut bear design.

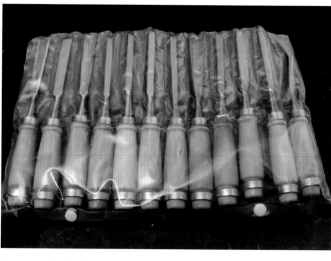

An example of a chisel set, but you can use any type of flat chisel.

Rasps.

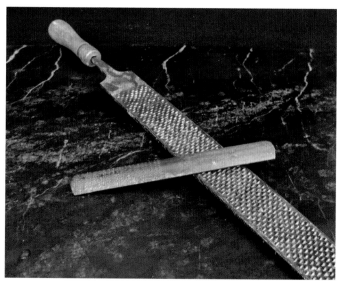

Flat four sided file, large and small. For this project, we used the small file.

Spray lacquer, sealer, and polish.

Small box, old towel, old tray, and newspaper.

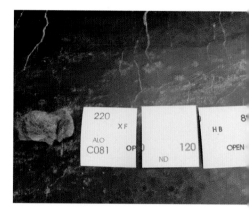

80-, 120-, and 220-grit sandpaper and 0000 steel wool.

You can use the project pattern or draw the design freehand. In this case, we drew the design freehand on the stone. At any time, during the creation of your sculpture, we encourage you to look ahead at the photos so that you can visually see the process before you execute any of the steps.

Take your precut piece of soapstone and place it in your hands so that the head and nose are facing to the right. Pick up your pencil and begin drawing the design.

Take your piece and begin with a center line down the middle of the top.

Continue down the chest and paws.

Make sure the line goes all the way to the edges. (Go back over the line to make it darker if needed.)

Turn your piece back head side up and begin drawing circles for the ear placement.

Face the bear towards you and continue the center line down the middle of the face.

Draw a line from one edge to the other, separating the ears from the back of the bear.

In a side view, extend the line down slightly.
This line represents the height of the ear.

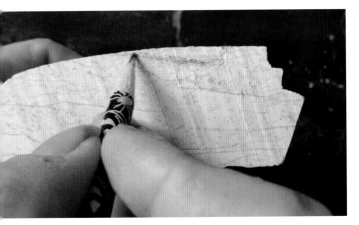

At the desired depth, begin drawing a straight line to the
middle of the back. These lines represent the shape of your
bear. Keep in mind this is a stylized bear that is very round.

Depending on your pre-blocked bear, you may or may not want to
remove additional material from the chest area. In our case, we did.
We extended the curve to match the curve of the back of the bear.
The line goes all the way to the top of the paws.

Here, you can either use the patterns provided or draw in the lines of
the legs freehand. We chose to draw them freehand. The line you are
going to create here represents the placement of the hind leg. Begin
midway from the top of the back and extend the line down to match
the height of the front paw.

Once the height matches up, draw in the back paw.

Now go back up to the top and draw in the crease of the front leg.

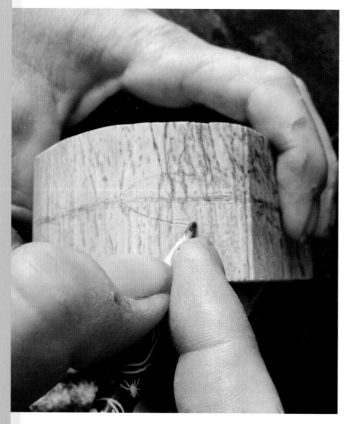

Turn the piece around and begin drawing in the tail. Again, you can use the pattern or draw in the tail freehand as we have.

Turn the piece around and repeat the ear placement line on the other side.

Once the desired ear depth is reached (should match the other side), begin drawing the line for the top of the back. Remember that these lines create the shape of the bear.

Turn the piece to face the front and begin drawing in the height of the paws.

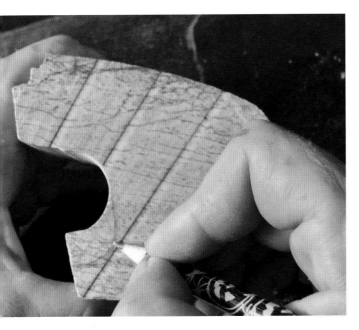

Because we are not satisfied with the blocked piece as it exists, we are drawing the line of material to be removed as we did on the other side.

Turn the piece towards you and draw in the nose lines. (Make the lines darker if needed.)

Now go back to the head and begin drawing in the shape of the snout. Be mindful of the center line.

Below the nose, draw in a line straight across from side to side.

Repeat on the other side.

Continue the line onto the side of the piece.

Draw in the brow line.

This line, once again, represents the shape of the head and chest, showing material that has to be removed.

Draw in the ear line, as with the other side.

Use a flat chisel to remove material to define the shape of the sitting bear. If you do not have a flat chisel, you can use any type of rasp. The chisel, however, is recommended.

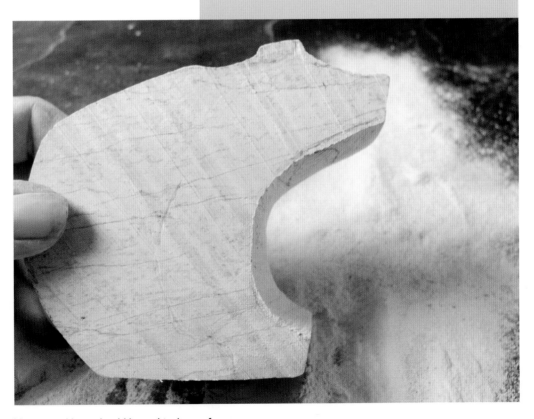

Your round bear should have this shape after removing the material as shown in this chapter.

To begin removing material, start on the back behind the ear. Use your flat file; be sure the rounded side is down. Place it on the line that you created behind the ears and, in a sweeping motion, remove material to create an indentation (as shown in the next photo).

Once you have filed to the desired depth, your bear's silhouette should look something like this. Don't go too deep. Remember, it is always better to remove less material at first.

Now place the rounded side of the file down on the line in front of the ears. You are encouraged to look ahead at the next photo for guidance. Using the same sweeping motion, file material off the top of the head to create a sloping snout.

Stop frequently so as to not go too deep.

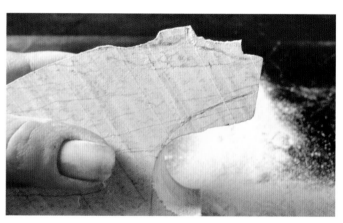

Here you can see that the first step is removed and there is still some material to go.

Continue filing the snout until you reach the desired profile.

Now turn the piece to where it is comfortable for you to file the chest area. With the flat side of the file, begin removing material under the jaw following your line.

Begin filing.

Here the material has been removed to the line. Be sure not to file too much; stop frequently to evaluate your depth.

Your bear should look something like this.

Now shift your attention to the paw area.

Creating the Legs

Now we're going to create the legs and define the paws.

The crevice will define the legs. Be sure not to go too deep at this time.

Begin by going back over the lines to darken if necessary. In this case, we went over all the lines on this side of the bear.

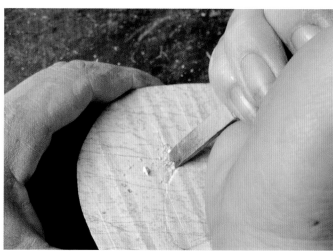

Once the lines are etched, take your chisel and, in a twisting motion, pivot the chisel on its tip to remove material.

Now get a small flat chisel and, with the tip of the chisel, trace over the lines to begin creating the crevice.

Go slowly and do not apply too much pressure at first.

At this point, you should have a teardrop-shaped indentation. Be sure to look ahead at any time to visualize the next step.

Step back and take a look at your piece.

Continue the motion down the legs…

Now go back over your lines, adding depth to the crevice.

…and down to the edge of the piece.

Continue working the lines. The twisting motion is particularly effective.

The twisting technique close-up looks like this.
The chisel is planted on the line.

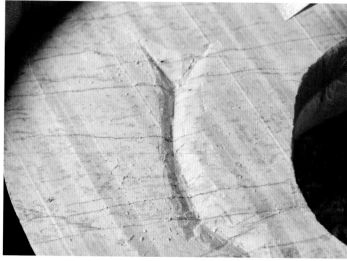

The shape begins to emerge.

The pressure is applied to the left corner, digging into
the material on the left and twisting to the left.

Work on any areas needed, in this case the paws.

The chisel pivots on itself and ends up on the line again.
A thin layer of stone has been removed.

Pick up your rasp to work with the front paw.

If you do not have a rasp, you can use a flat file.

The definition of the back leg should begin to show.

In a sweeping motion, remove material from the front paws, to create the depth you desire.

For the contour of the paw, go back to your small flat chisel.

Be sure to hug the line as to not file down the back leg.

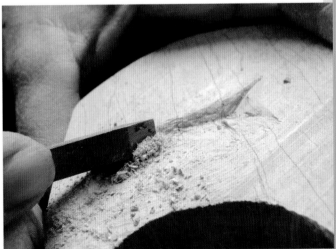

Here we are using a technique where the chisel is vertical against the back leg. Deepen the crevice by following the contour of the leg. The point of the chisel is well planted into the stone. Do not apply too much pressure; work slowly.

The crevice deepens creating the shape of the legs.

... as well as the crease.

The small flat chisel can also be used to round the shape of the legs...

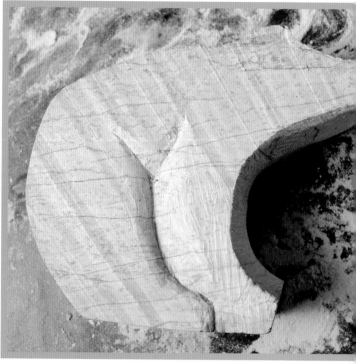

At this point your bear should look something like this.

Rounding the Body

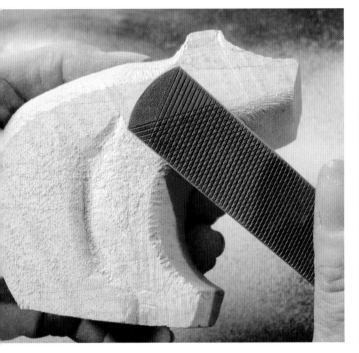

Rounding the body.

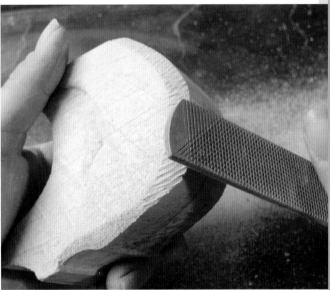

Follow the edge down the back.

Pick up the bear and your flat file. With the flat side up begin rounding the edges.

Make sure that you do not apply too much pressure, let the motion remove the stone. Be careful when working around the ears.

We started on the top of the bear.

Frequently pause to take a look at the shape that you are creating.

At this point, a view from the top, we are switching to the other side. Look at the ears and the shape that is beginning to emerge.

At this stage, the bear could use more rounding. The edge is filed off, but the bear still has a blocky look. In an upward motion, continue rounding the body.

File down to the edge of the piece.

Begin rounding the chest area. At this stage, you do not want to use a sawing motion on the neck, rather a twisting motion where the file travels vertically on the chest.

File once again, gently.

Following the edges,
continue to round the bear.

Take a good look at the other side for symmetry. You do not want to remove too much material at this stage.

Gently round the snout.

Creating the Head

Creating the head.

Before removing material, go back over the lines with your pencil.

Once your lines are darkened, begin removing material on the outside of the snout. In this case, we started on the right side.

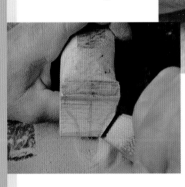

Using the flat side of the file, in a sawing motion, watch as the material is removed. Do it slowly and stay to the outside of the darkened line.

Once you have reached the outside of the line, flip the file over. Using the rounded side of the file gently file the curve of the snout.

It is important to periodically check the shape you are creating.

Once you are satisfied with the initial shape, begin filing on the other side. First with the flat side of the file…

…then the rounded side.

Pause, take a look at the shape you have created…

…making sure that the depth is even on both sides. Be sure to check your center line.

Now begin rounding the edges of the snout. Here we started on the right.

Follow the edges. You will be using the rounded side of the file.

Go back to the head... Where the snout begins to "Y" (widen), draw a horizontal line from one edge of the snout to the other. Feel free to look ahead at the photos for clarity.

This will be your guide to defining the eyebrows.

Now pick up a rasp and gently file across the line. Be careful not to apply too much pressure, you are only creating a ridge.

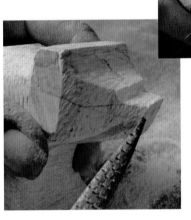

The ridge should look something like this.

The bear in its entirety should look something like this.

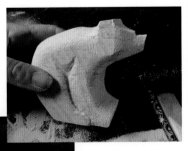

Gently round the tip of the snout with the rasp. With the tip of the rasp, you can shape the snout with more detail.

Very gently, with the tip of the rasp, create a slight angle on the edge of the snout. Look ahead at the next few photos. This is achieved in just a few motions.

Note the slight angle on the rounded snout.

Following that angle, work gently on the side of the head.

Look at how the shape begins to emerge. Do not apply too much pressure and work slowly.

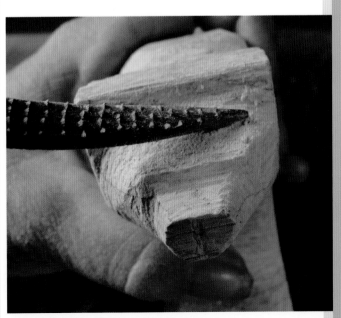

Now begin on the other side.

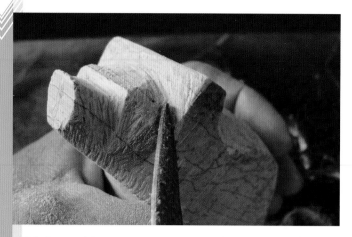

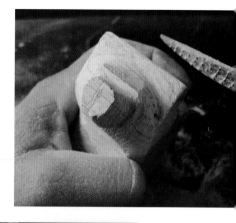

Pause to take a look at your piece. Notice any areas that need to be worked on.

With the tip of the rasp, create a slight indentation at the base of the ears following the rounded shape of the head.

Here we are further defining the snout.

Gently remove some material.

Gently round the snout area.

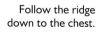

Follow the ridge down to the chest.

Your bear should look something like this.

Rounding the Flat Side of the Bear

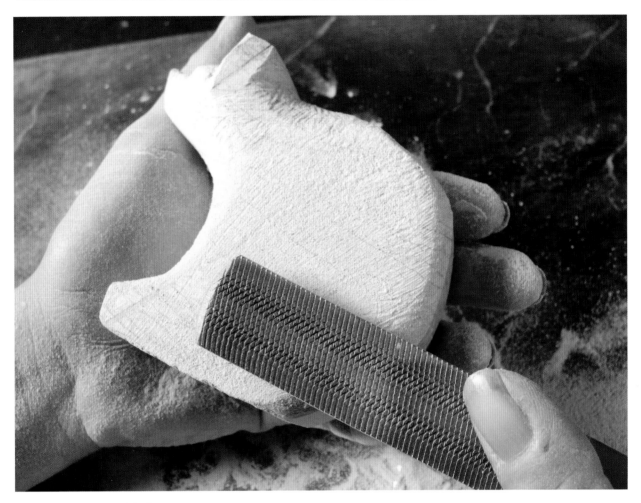

Rounding the flat side of the bear.

Here, we are rounding the flat side of the bear, not choosing to create leg definition. However, it is your choice: if you would like to create legs on both sides of your bear, review "Creating the Legs." If you have already created legs on both sides, continue rounding the whole body of the bear, but be mindful to leave the tail area and the top of the back alone. These areas will be addressed in "Preparing the Tail."

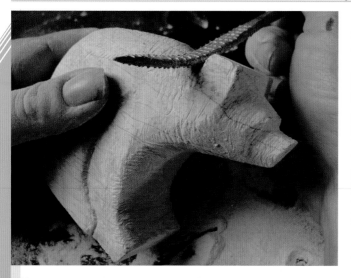

Take your rasp and remove material behind the ear.

Sweep the rasp down to the paws. Again, do this in several motions, working slowly.

Work slowly and don't apply too much pressure. It's better to go over the area several times than remove too much material.

Now that the contour of the chest, neck, and ear are defined, begin shaping the remainder of the bear using the round side of the flat file.

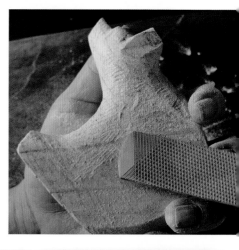

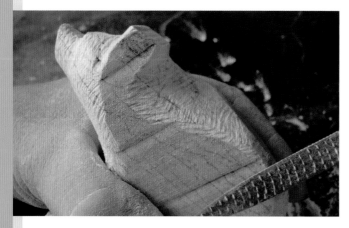

You can always take away more, but you cannot put it back. Work slowly to create the rounding of the edges.

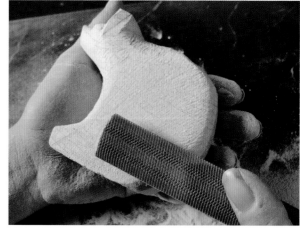

Notice that the saw marks from the blocked piece have been removed by the filing.

Continue rounding beneath the ear, removing the flat material.

Here we were not satisfied with the shape of the bear behind the ears, so we are rounding that area a bit more.

Preparing the Tail

Preparing the tail.

Now, we will focus on defining the tail of your bear, as well as the sloping of the upper back above the tail. Notice the ridgeline that runs from the tip of the tail towards the head. This is a very subtle detail, but it is there.

Use the flat file, rounded side down, and begin removing material working towards the center line. Do not go over the center line.

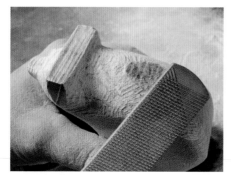

Begin creating a crevice on the outside of the tail line using the chisel in a twisting motion. This is not going to be a straight line, but a curved one. Feel free to look ahead.

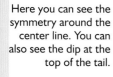

Try to work both sides evenly, coming in a little bit at the top of the tail.

The chisel pivots out to the left from the right corner.

Here you can see the symmetry around the center line. You can also see the dip at the top of the tail.

As you can see here, the result of the twisting motion is well-defined. Be sure to stay on the outside of your tail line.

In the following pictures, we will be outlining around the tail, using a flat chisel in a twisting motion. Start at the centerline at the end of the tail. Rock the chisel into the stone.

Repeat the motion up the tail line, creating your crevice.

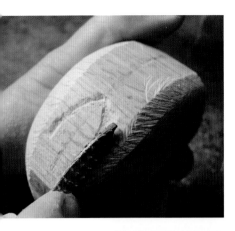

Once your tail is outlined with the chisel, get the rasp and begin removing material around the crevice.

Removing material around the tail crevice will eventually make the tail three-dimensional. This is called subtractive sculpting to achieve a bar relief look to the tail.

Be sure to stay away from the tail and the centerline should your bear need work around the hips.

Here the tail appears to emerge.

Now that the definition of the tail is emerging, we noticed that the hips needed more rounding.

Now that we are satisfied with the definition of the hips, notice the ridge beginning to appear above the tail.

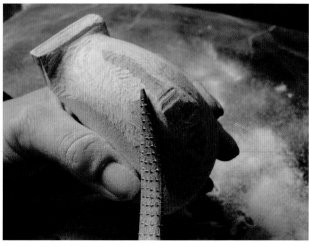

Turn the piece top side up and work around the ridge.

The more material you remove around the tail, the more prominent it will be. We chose to use the rasp in a sweeping motion to further define our tail.

With the rasp, in a sweeping motion, gently remove material.

Step back and take a look at your piece.
Decide if you want to remove more material.

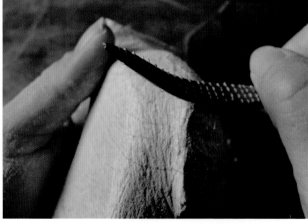

Here is where the ridge of the bear meets the tail.

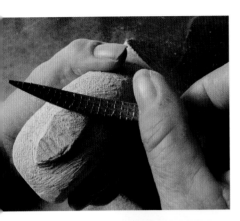

Brush the rasp in a light motion down the tail.

In this case, we are gently rounding the sides of the tail.

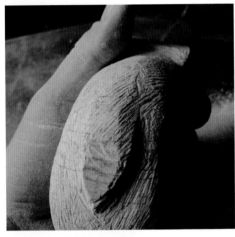

At this point your piece should look something like this.

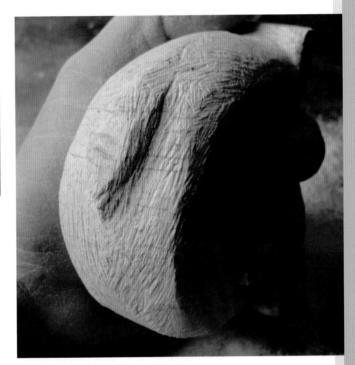

With the tip of the rasp, slope the ridge delicately. Don't be too aggressive, this is a small and subtle movement.

Notice how the tail seems more pronounced.

At this point, if you are not satisfied with the shape of your tail, you may wish to work on it.

Here we are working on the shape of the bear.

You can also use a flat file to further round the bear.

Work slowly and gently... This is a scraping motion with the chisel barely removing material in one sweep.

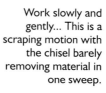

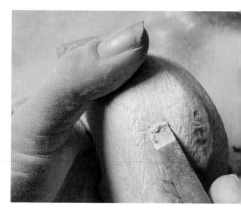

We decided after the shaping of the body, to further define the tail using the flat chisel.

In a scraping motion, follow the curve of the tail upward.

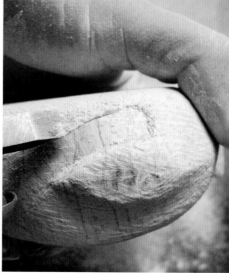

Notice how subtle the removal of material is.

Then repeat... Look ahead for clarity.

Repeat on the other side if needed.

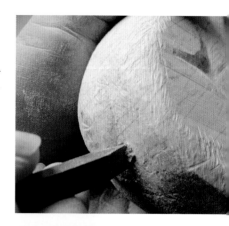

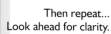

Now we are using the flat chisel in almost a shovel motion along the tail line. The goal here is to remove material so that the depth is even on both sides.

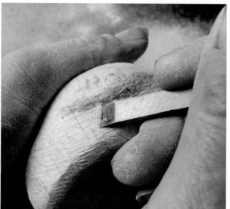

Again, this is a subtle movement.

Your bear should look something like this.

Creating the Ears

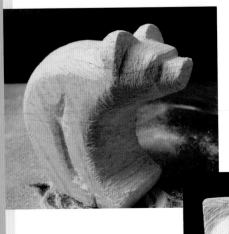

Work in stages...
Now we go a little deeper.

Creating the ears.

Here you can see the groove widen and we have reached our desired depth. Don't go too deep into the forehead.

Start with the bear facing towards you.

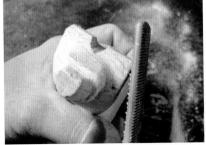

Now pick up your flat file and begin rounding the corner of the ear. Be sure to have a firm grip on your bear. Once again, do not apply too much pressure.

Pick up your rasp and place it on the center line of your piece.

With a few sweeps, your ear should look like this.

Gently notch out a groove between the ears. Do not remove too much at this stage. The motion used here to remove the material is a sawing "back and forth" motion. You do not need to apply a lot of pressure as the movement removes the stone.

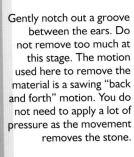

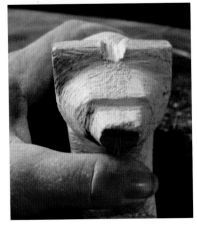

Now file the other side.

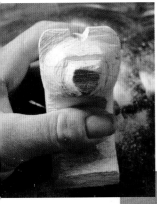

Both exterior sides of the ears should be rounded.

The tip of the rasp is especially useful for the interior of the ear.

Now with the rasp, begin gently rounding the interior and back of the ear. Go slowly as this area is delicate.

In a circular motion, smooth out the area where the ear meets the back.

You can play with different techniques. If you hold the rasp closer to the tip, you can remove material by rolling the rasp against the stone.

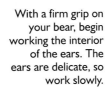

With a firm grip on your bear, begin working the interior of the ears. The ears are delicate, so work slowly.

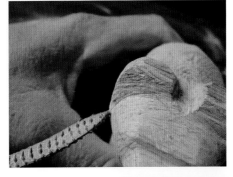

With the tip of the rasp, create a circular line at the base of the ear, defining the ear from the head.

Continue working behind the ear.

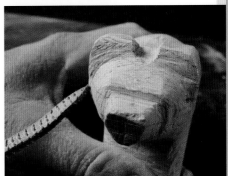

This is a subtle detail, but notice the rounding of the head where the ear meets the forehead.

It is a good idea to pause and evaluate where you are. Take a look at the back of your ears and notice the difference between the two.

Front view. Your piece should look something like this. Notice how symmetrical the ears are.

View from the front... Notice the dramatic difference between the ear on the left and the ear on the right.

A view from a different angle... Your ears should look something like this.

Other side... Notice the balance.

Take your rasp and begin working on the other ear. Work slowly, this is a delicate area. Make sure you have a firm grip on your piece. Round in a soft circular motion.

Work on rounding the inside of the ears maintaining the slope of the ear.

Back view.

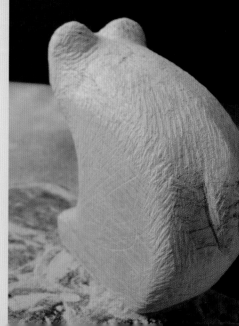

This is the final stage of your project and you will need a small file, sandpaper, and some lacquer finish (or whatever you wish to use to polish your piece). We did not use steel wool in this project. However, it is commonly used after sanding and we suggest 0000 steel wool should you choose to use it. We always say this stage "...separates the men from the boys..." — people who spend the time properly sanding will reap the rewards of what the piece will look like when it is polished. We use 80-, 120-, and 220-grit sandpaper in this project, but you can go up to 800-grit if you choose; the finer the grit, the smoother the polish. This stage of the project is perhaps the most important.

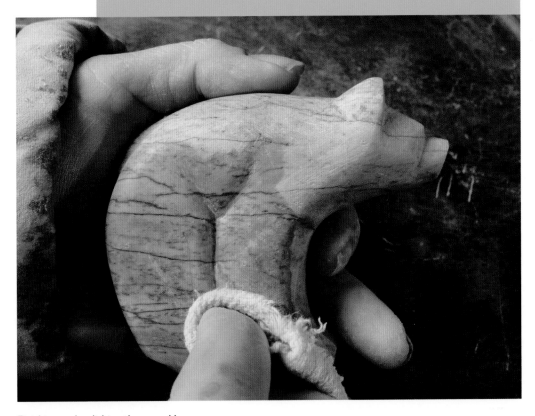

Finishing and polishing the round bear.

You will need sandpaper and a flat file. Folding the sandpaper in a triangle allows you to work the corners and creases easily.

Begin anywhere. We began on the back...

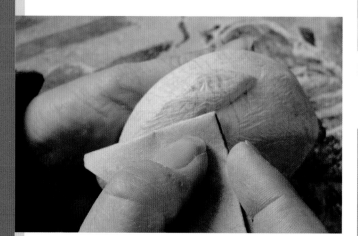

Here we are smoothing out the tail area.

... and then moved to the chest area.

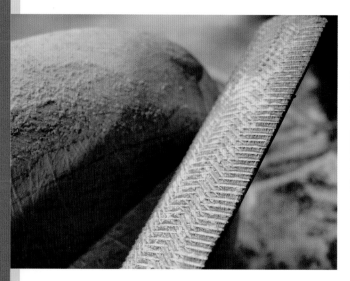

However, you can use the fine side of the flat file to smooth your piece. This process speeds things up for the removal of the rasp marks.

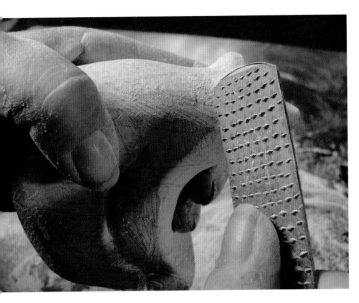

When smoothing behind the ears, be sure to have a good grip on the piece and work slowly.

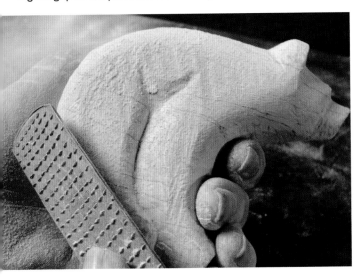

Continue smoothing in a slow, brushing motion, not applying too much pressure.

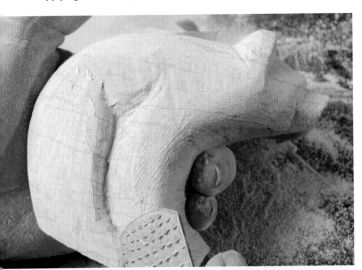

Here you can see the piece begin to have a smoother finish.

As the surface smoothens, notice how the shape of the bear shows more detail, such as the tail area.

Here we are not sanding but rather gently smoothing the facial area.

Use a gentle touch behind the ears as well.

In a wiping motion, sand the side of the bear. Watch as the file marks disappear.

Notice how you can still see some of the file marks, but the surface is just a little smoother. These file marks will be removed when we begin using the sandpaper.

Sand the tail area.

Sand the body and the legs.

Before sanding, take a look at the bear and work on any details or changes you deem necessary.

Once you have sanded the whole body with the 80 grit sandpaper, sand the whole piece with the 120 grit...

First begin with the 80 grit, your coarsest sandpaper.

...and then the 220 grit sandpaper.

At this stage, you could begin a sanding technique called "Wet-Dry" sanding. This technique involves sanding the piece while wet to achieve a much smoother surface.

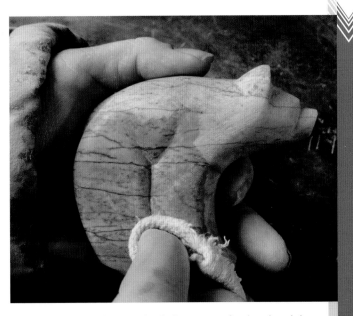

We chose to continue with the dry technique. Here we are smoothing out the last stage of our project — the head area — with 220-grit.

Allow the piece to dry completely (minimum of an hour) and then begin applying the polish. In this case, we applied a mixture of oil and beeswax to the piece with a soft cloth... See how the piece comes alive with color.

Pause, step back and make sure that you are satisfied with the smoothness of your bear. Look carefully around the whole bear.

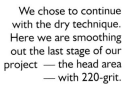

We suggest you clean the entire surface to remove all dust and residue from the bear. In this case, we used water. You can wash it under the tap or use a cloth; tap is recommended. When the piece is wet you get a preview of what the piece will look like when polished.

Congratulations! You have finished your round, sitting bear!

Sitting Bear Patterns

Patterns to use on a block or raw piece of soapstone. We recommend that you make copies of the patterns before cutting them out.

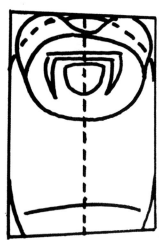

Front view of the round bear pattern.

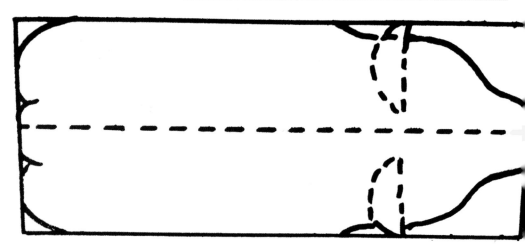

Top view of the round bear pattern.

Back view of the round bear pattern.

Side view of the round bear pattern.